Anonymous

Report of the Autopsy of the Siamese Twins

Together with other interesting information concerning their life

Anonymous

Report of the Autopsy of the Siamese Twins
Together with other interesting information concerning their life

ISBN/EAN: 9783337254599

Printed in Europe, USA, Canada, Australia, Japan

Cover: Foto ©berggeist007 / pixelio.de

More available books at **www.hansebooks.com**

REPORT OF THE AUTOPSY

OF

THE SIAMESE TWINS,

TOGETHER WITH

OTHER INTERESTING INFORMATION CONCERNING
THEIR LIFE.

(REPRINTED FROM THE PHILADELPHIA MEDICAL TIMES.)

PHILADELPHIA:

J. B. LIPPINCOTT & CO.

1874.

THE SIAMESE TWINS.

OF

THE SIAMESE TWINS,

TOGETHER WITH

OTHER INTERESTING INFORMATION CONCERNING THEIR LIFE.

(REPRINTED FROM THE PHILADELPHIA MEDICAL TIMES.)

PHILADELPHIA:

J. B. LIPPINCOTT & CO.

1874.

LIPPINCOTT'S PRESS, PHILADELPHIA.

REPORT OF THE AUTOPSY

OF

THE SIAMESE TWINS.

A SKETCH OF THE LIFE OF CHANG AND ENG.

IN the present article it is not purposed to give an elaborate history of these famous twins, but only to put on record certain well-ascertained facts of physiological interest in regard to their life, to give all that can be ascertained as to the circumstances of their death, and to offer a brief history of the manner in which their bodies were brought to Philadelphia.

According to a writer in *Lippincott's Magazine* for March, the Siamese Twins were born in 1811, some thirty miles southwest of Bangkok; their father being a Chinaman, their mother a native of Siam, bred by a Chinese father. The twins were, therefore, three-fourths Chinese, and were known in their native home as the "*Chinese* Twins." They were the first-born sons of their parents; but their mother has presented her husband with four other pairs of twins

5

and four children born at single births,—all of them
normal and healthy. Their mother, during their
infancy, entirely recognized their separate individu-
ality, and also the fact that there existed a common
sensibility in the centre of the band. She stated,
what is undoubtedly true, that at first the ligament
was so short that the boys were compelled always to
be face to face; even in the bed they could not turn
without being lifted up and placed in the desired
position. As they grew, the ligament seemed grad-
ually to stretch, until they were able to stand side by
side, and even back to back, and to turn themselves
in bed by rolling one over the other. The father
being a fisherman of the laboring class, the boys
lived in one of the floating houses of the country,
and soon became famous swimmers, spending
much of their time in the river. It was the pecu-
liarity of their movements in the water which first
attracted the attention of a Scotch merchant, Mr.
Robert Hunter, and finally led to their leaving their
native country in quest of fortune.

For many years Chang and Eng Bunker lived in
North Carolina, where they were married, and raised
large families of children, Chang being the father
of ten, Eng of nine. Dr. Joseph Hollingsworth, to
whom we are indebted for the information given
in this article as to their habits of life and the
circumstances of their death, states that he has
known them as residents in the neighborhood of
Mount Airy, North Carolina, for some twenty years,
during which time he has acted as their family phy-

sician. Chang, who is said to have derived his name
from the Siamese word for "left," was the left of the
pair, and was much smaller and more feeble than
his brother Eng, whose name signifies "right."
Their habits were very active : during the latter
part of their life they and their families lived in
two houses, about a mile and a half apart, and it
was an inflexible rule that they should pass three
days alternately at each house. So determinedly
was this alternation maintained that sickness and
death in one family had no effect upon the movements of the father, and a dying or dead child was
on one occasion left in obedience to it : indeed,
Dr. Hollingsworth is very positively of the opinion
that the death of the twins themselves was the result
of this rule, or, at least, was materially hastened
thereby. This will be made apparent hereafter.

The evidences during life that the twins were
physiologically distinct entities were very numerous
and apparent. They were different in form, tastes,
and disposition ; all their physical functions were
performed separately and unconnectedly. What
Chang liked to eat, Eng detested. Eng was very
good-natured, Chang cross and irritable. The
sickness of one had no effect upon the other, so that
while one would be suffering from fever the pulse of
the other would beat at its natural rate. The twins
not rarely suffered from bilious attacks, but one
never suffered at the same time as the other ; a cir-
cumstance which seems somewhat singular in view
of the close connection which the post-mortem

has shown to exist between the livers of the pair.
Chang drank pretty heavily,—at times getting drunk;
but Eng never felt any influence from the debauch
of his brother,—a seemingly conclusive proof that
there was no free interchange in their circulations.

The twins often quarrelled; and of course, under
the circumstances, their quarrels were bitter. They
sometimes came to blows, and on one occasion came
under the jurisdiction of the courts. After one of
these difficulties Chang and Eng applied to Dr. Hollingsworth to separate them, stating that they could
not live longer together. Eng affirmed that Chang
was so bad that he could live no longer with him;
and Chang stated that he was satisfied to be separated, only asking that he be given an equal chance
with his brother, and that the band be cut exactly
in the middle. But as Dr. Hollingsworth advised
very decidedly against this, and declined to inter-
fere, cooler counsels prevailed.

In August, 1870, Chang suffered from a paralytic
stroke, from which he never fully recovered; and
during the last year of his life he several times said
to Dr. Hollingsworth, "*We* can't live long."

On the Thursday evening preceding their death,
the time having arrived for their departure from
the house at which they were staying, the twins
rode a mile and a half in an open wagon. The
weather was very cold,—the night being the severest
of the winter. Chang had been complaining for
some days of cough, with distress and actual pain
in the chest. He was so unwell that his wife thought

he would be unable to bear the trip; but he finally went. On Friday morning Chang reported that he felt better, but that in the night he had had such severe pain in the chest, and so much distress, that he thought he should have died.

The twins slept in a room by themselves or with only a very young child present; and some time in the course of Friday night they got up and sat by the fire. As they were accustomed to do this frequently, nothing was thought of it by those of the family who saw them, even though they heard Eng saying he was sleepy and wanted to retire, and Chang insisting on remaining up, stating that his breathing was so bad that it would kill him to lie down. Finally, however, the couple went to bed again, and after an hour or so the family heard some one call. No one went to the twins for some little time, and, when they did go, Chang was dead, and Eng was awake. He told his wife that he was very "bad off," and could not live. He complained of agonizing pain and distress, especially in his limbs. His surface was covered with a cold sweat. At his request his wife and children rubbed his legs and arms, and pulled and stretched them forcibly. This was steadily continued until he went into a stupor, which took place about an hour after the family were alarmed. The stupor continued up to death: according to the statements of the family, there were no convulsions.

Dr. Hollingsworth did not reach the house until after the death of both of the twins. He found the wives, and especially the children, averse to any post-mortem being made, but, after much persuasion, obtained permission to put the bodies in a position to be preserved until he could obtain some one from Philadelphia to perform the autopsy. He placed the bodies, after they had been thoroughly cooled, in a coffin, which was put in a wooden box, which was, in its turn, enclosed in tin; the whole being buried in a dry cellar in such a way as to be imbedded in charcoal.

As bearing upon the question, What was the cause of the death of Chang? it is important to state that Dr. Hollingsworth had repeatedly told Chang and Eng that, in his opinion, the death of one did not necessarily compromise the life of the other; that he could separate them, by cutting close to the body of the dead one, without killing the living one. It would appear possible, in view of this, that the death of Eng was not simply the result of fright.

———

SHORTLY after the death of the Siamese Twins, Dr. William Pancoast requested the Mayor of Philadelphia to telegraph to the Mayor of Greensboro', North Carolina, in regard to the possibility of a post-mortem examination being obtained. To this the Mayor of Greensboro' substantially replied that he had neither knowledge nor power in the matter. When Dr. Hollingsworth, *en route* North, arrived at Greensboro', the telegram of Dr. Pancoast was handed to him. On the evening of his arrival

at Philadelphia (Friday) he saw Dr. Pancoast and
Prof. Gross, and a letter was written to the wives
of the twins, proposing that Dr. Pancoast should
come on to embalm and examine the bodies.

On Sunday Dr. Hollingsworth saw Prof. John Neill,
and, on consultation, it was concluded that the
matter was of public importance, and should not be
confined to any single private individual. As the
promptest method, it was deemed best to call a
meeting of such physicians as were interested in the
matter and could be hastily got together.

The meeting took place on the evening of Monday, January 26, 1874, at the house of Dr. Neill;
but, although a number had been asked, only Prof.
Leidy and Dr. Ruschenberger, besides Drs. Hollingsworth and Neill, were present at the conference.
As the result of their deliberations, it was deter-
mined that two gentlemen should be at once dis-
patched to the homes of the twins, in order to
examine and embalm the bodies as speedily as possible; and it was agreed that Drs. William H. Pancoast and Harrison Allen should be requested to go.

It will be seen at once that the College of Physicians was in no wise responsible for the acts of
the Commission, although its name was freely used
by the prominent Fellows engaged in the transaction.
Indeed, these gentlemen, recognizing this, were
prepared to meet the expenses of the trip had the
College failed to endorse their action.

Owing to various obstacles and embarrassments, the
Commission did not leave the city until Thursday

night, January 29. At the request of Dr. Pancoast,
Dr. Andrews went with the party as a companion
and aid.

The Commission arrived at Mt. Airy on the evening of Saturday, January 31, and proceeded to the
residence of Eng the following morning, in com-
pany with a photographer and Dr. William Hollingsworth, who is the family physician in the absence of Dr. Joseph Hollingsworth. The widows
of the twins received the Commission hospitably,
and a conference was at once entered into, at which
the "Mistresses" Bunker, the Commission, Dr.
Hollingsworth, and the widows' legal adviser were
present. It was then agreed that, under consideration of embalming the bodies of the twins, permission would be granted to exhume and examine the
structures distinguishing them, provided that no
incisions should be made which would impair the external surface of the band. Subsequently it was agreed
that limited incisions would be allowed on the posterior surface of the band. An agreement in writ-
ing was then drawn up, expressing the above restric-
tions, but extending authority to the Commission
to remove the bodies to Philadelphia, provided that
they be kept there in a fire-proof building, and held
subject to the commands of the families when informed of the completion of the embalming process.

The object of the visit of the Commission, having
been noised about the country, had attracted a
crowd of curious people, who were willing enough

to give the necessary aid in exhuming the bodies. The circumstances attending this were briefly as follows. The bodies were buried in the cellar of Eng's house, in a shallow grave, which had been covered with a tumulus of powdered charcoal. This being removed revealed several planks covering an outer wooden box, which, in turn, enclosed a tin encasement to the coffin. After unsoldering the tin box, the coffin was carried to the second floor of the house, to a large chamber. The lid was unscrewed, and the object of the search of the Commission was exposed to view. It was certainly an anxious moment. Fifteen days had elapsed since death, and no preservative had been employed. It was an agreeable surprise, therefore, that no odor of decomposition escaped into the room, and that the features gave no evidence of impending decay. On the contrary, the face of Eng was that of one sleeping; and the only unfavorable appearance in Chang was a slight lividity of the lips and a purplish discoloration about the ears. The widows at this point entered the room, and, amid the respectful silence of all present, took a last look at the remains.

The room was then cleared of all not connected with the work of the Commission; the bodies were disrobed, and preparations at once begun to secure photographs. The bodies were held in an erect position, and negatives of the entire figures, and views of the band at short foci, were secured. The day being cloudy, much time was necessarily expended in obtaining these pictures,—time sufficient for a number of observations upon the external appearance of the bodies to be recorded. The notes are given just as they were taken at the time:

Examination made Sunday, February 1, 1874, *fifteen days and eight hours after the death of Chang.*

The bodies were found in the coffin in a good state of preservation; there was a slight cadaveric odor about Chang, with marked passive congestion of the back of the arms and neck on both sides, and in a less degree of the posterior aspect of the forearms, buttocks, thighs, and legs; there was none on the feet, where, however, there was marked fulness of the superficial veins; this was better marked on the left side. There was a greenish discoloration on the anterior abdominal wall.

Face.—Lips moist and discolored; peculiar reddish congestion sparsely distributed over malar prominence and beneath ear. The thoracic discoloration was much deeper on the side next to Eng.

The left nipple was visible in front well towards the middle line, the right one just showing. The fingers of the right hand—the paralyzed side—were forcibly flexed, although *rigor mortis* was absent.

In Eng there was passive congestion of back, most marked on buttocks and infra-spinous spaces—none on the front of the body; slight greenish discoloration of anterior abdominal wall.

In both subjects the hair of the head was gray.

On *the pubis* of each subject the hair of the *left side* was *gray*, that of the *right side*, *black*.

The process of embalming was now begun. Incisions were made to the outer side of the median line of the abdomen in each individual, extending from the inferior margin of the thorax to a point midway between the symphysis pubis and the anterior superior spinous process of the ilium. The aorta was reached after the usual method, but was found to be in an atheromatous condition, compelling the selection of the left primitive iliac for the insertion of the pipe. A solution of chloride of zinc was then injected.

After the embalmment had been completed, the incision was continued upward and inward towards the band. Examination of the band through this incision convinced the Commission of the complex nature of the band, and suggested the suspension of a complete study of the parts until removal of the bodies to Philadelphia. The fact that the photograph had been far from satisfactory strengthened the Commission in its decision to stop the investigation at this stage. The incisions were, therefore, sewn up, the clothing readjusted, and the bodies placed in the coffin and conveyed to Mt. Airy. Here the tin box which was used for the temporary burial was again brought into use, and the lid carefully resoldered. Without delay the Commission started on its return, expressing the bodies at Salem.

The Commission arrived in Philadelphia, February 5, having been absent one week.

Upon the arrival of the bodies at the College of Physicians, they were placed in the care of the committee upon the Mütter Museum and of the Hall Committee, and were closely locked and guarded until a special meeting of the College was called, upon Monday evening, February 8, when, after considerable discussion, it was agreed that the College should accept the action of its Fellows and pay the expenses of the trip. Further, a vote of thanks was given to the gentlemen who went to North Carolina, and to Dr. Hollingsworth, and the Mütter Committee was authorized to appropriate three hundred and fifty dollars for the preparation of casts and photographs, which should remain the property of the Museum. Finally, the College appointed the Mütter Museum Committee and the original Commission (Drs. Pancoast and Allen) as a joint committee for carrying out the examination of the Siamese Twins; it being understood that a report and a demonstration of the specimens were to be made to a subsequent meeting of the College; also, that the dissections and the report were to be the work of the original Commission.

On Tuesday, the 10th instant, they were exposed for study: they were at that time found in a satisfactory condition, except the right lower extremity of Chang, which required immediate care to prevent further destructive changes taking place.

STATEMENT.

Statement of Eng's widow, made to the Commission at Mount Airy.—The paralytic stroke from which Chang had suffered occurred about three years ago, when he was at sea, seven days out from Liverpool. He had been intemperate for some time previously, and had been drinking hard on board the vessel, being frequently intoxicated. He had never had mania à potu. Even when he was drunk, Eng was not affected. Two of his children had died; one from phthisis, the other apparently from apoplexy. Eng had lost five children: one each from phthisis, diphtheria, and dysentery, one from the effects of a burn, and one still-born.

Chang died Saturday, January 17. He had had a cough since the preceding Monday night. It was dry, with scanty, frothy sputum and no pain. Left lung probably involved; slight dulness on corresponding side. On Thursday, January 15, his skin was acting freely, and for that reason Dr. Hollingsworth ordered him not to venture out; but, in spite of that prohibition, he went as usual to Eng's. At the time of his arrival he had little cough and no expectoration, but loud bronchial râles were plainly heard by those around him.

When Eng saw his wife after learning that Chang was dead, he said, "I am dying," but did not speak of his brother's death. He soon afterwards expressed a desire to defecate, and this continued for half an hour. He rubbed his upper extremities, raised them restlessly, and complained of a choking sensation. The only notice he took of Chang was to move him nearer. His last words were, "May the Lord have mercy on my soul!"

PHONOGRAPHIC REPORT.

THE SIAMESE TWINS AT THE COLLEGE OF PHYSICIANS.

A SPECIAL meeting of the College of Physicians of Philadelphia was held at the hall, Wednesday evening, February 18, for the purpose of hearing the report of the Commission on the Siamese Twins,—Dr. W. S. W. Ruschenberger, U.S.N., in the chair. On motion of Dr. Gross, it was, after some discussion, resolved that the *Philadelphia Medical Times* be allowed to report the proceedings of the meeting, on condition that three hundred copies of the journal should be given to the college for the use of the members.

The bodies of the Siamese Twins being upon the table, the meeting proceeded to hear the report of Drs. Pancoast and Allen. On behalf of the Commission, Dr. Pancoast stated that, the dissection not having been entirely completed, their report would be a verbal one, to be followed at some later date by a memoir upon the subject. He further remarked that it had been agreed that he should consider chiefly the surgical aspect of the matter in hand, whilst to his colleague had been assigned the demonstration of the anatomical peculiarities.

DR. WILLIAM H. PANCOAST:

Mr. Chairman, and Fellows of the College :— Having been requested, as a member of the Com-

mission, to open the discussion this evening, I will
say briefly, in reference to this monster of a symmetrical duplex development, joined, as many of
the Fellows now know, at the ensiform appendix
and also here at the omphalos or navel, that at the
investigation which we made on the first occasion
at Mount Airy I made the opening incision of the
body on the line for the ligation of the primitive
iliac, on the right side; Dr. Allen made the incision
on the left. The object was to reach the great
vessels,—the aorta and two primitive iliacs,—and to
force the injecting material which we used for embalming (chloride of zinc) up the aorta and down
the iliacs until it ran from the incisions made in the
fingers and toes. It flowed freely through the blood-
vessels in Eng, owing to the ossified condition of
his arteries; the injection in Chang was, however,
not so successful, owing to decomposition in the
tissues and blood-vessels. It was necessary to re-
peat the injecting process several times in order to
preserve the body. The arteries of Chang were
found to be very much decomposed,—quite rotten,
in fact.

In Dunglison's Medical Dictionary we find the
scientific name given for the Siamese Twins, in the
classification of teratology, to be *Xiphopages;* and
by referring to the admirable article on Diplo-
teratology of Dr. G. J. Fisher (published in the
Transactions of the Medical Society of the State of
New York for the year 1866), it will be found that
the twins belong in the class of *Anacatadidyma.*

In his classification of double monsters he makes
three orders : *Order first*, — *Teratacatadidyma ;*
derived from τέρας, τέρατος, a "monster," κατά,
"down," and δίδυμος, a "twin." *Definition*,—duplicity, with more or less separation, of the cerebro-
spinal axis, from above downwards. *Order second*,
— *Terata-anadidyma*, derived from ανά, "up" or
"above," and δίδυμος, a "twin." *Definition*,—
duplicity, with more or less separation, of the
cerebro-spinal axis, from below upwards, or from
the caudal towards the cephalic extremity of the
neural axis. *Order third*,— *Terata-anacatadidyma*,
derived from ανά, "above," κατά, "down," and
δίδυμος, a "twin." *Definition*,—duplicity, with
more or less separation, of both the cephalic and
the caudal extremity of the cerebro-spinal axis,
existing contemporaneously. In this order, the
monster now before us might be called an *Omphalo-
xiphodidymus.*

Thus we have the scientific nomenclature of this
monster. Of course, the consideration of greatest
interest to the profession, and one of the main rea-
sons why the Commission made such exertions to
obtain this post-mortem, was that the American
profession might not be charged with having neg-
lected an effort to obtain an autopsy, which would
solve the mystery of their union. The feature of
greatest interest is connected with this band,—about
four inches long and eight inches in circumference.
In addition to this, there are other points of importance in teratology, in regard to the fulfilment

of the law of homologous union, in relation to the
juncture of the recti muscles and the fasciæ of the
obliquus and transversalis at their point of meeting
in the centre of the band. In regard to the position of the hearts, we think their apices present
towards each other. [This has since been found to
be the case.] The livers we have found to approxi-
mate to each other and to push through the respective
peritoneal openings into the band. We extended
our incisions to the margin of the band in front.
By placing my hand in the peritoneal cavity of
Eng and my colleague placing his hand in the peritoneal cavity of Chang, we pushed before us processes of peritoneum, which ran on to the median
line of the band ; and we could feel our fingers in
the lower portion of the band, behind the median
line, with a distinct layer of peritoneum between
them, demonstrating at once the prolongation of
the peritoneum into the band, and the complete
separation of one peritoneal cavity from the other
at this median line. Above that we felt some traces
of vascular connection, apparently running from
one liver to the other ; but this we will examine into
when we have a better opportunity of carefully dis-
secting and examining what vascular structures may
exist. We also noticed that in turning off the flaps
consisting of the anterior walls of the abdomen,
the hypogastric arteries, as illustrated by the diagram on the blackboard, ran upwards in each body
into the band. We lost them in this way, as
we think, towards the common umbilicus in the

anterior inferior surface of the middle of the band.

It is probable that the two hypogastric arteries on each side passed through this umbilicus. Whether or not there were two umbilical veins, we have not yet been able to decide, nor to answer the question whether the umbilical cord was double or single and composed of the four hypogastric arteries and two umbilical veins, or whether the placenta was single, double, or twin.

We also recognized that the ensiform appendix, as shown in the diagram of each side, was prolonged and united in the middle line. On our later examination, we find that there is complete continuity of structure of the cartilages, but no true joint at the middle line, although it is possible there may be some small synovial sacs farther up. The motion is mainly due, as I here demonstrate to you by moving these bodies one upon the other, to the elasticity of the connected ensiform appendices and intervening fibro-cartilages.

In regard to the vascular connection of the band, we have not yet been able to make so thorough and careful an examination as we wished; but still, in throwing colored plaster into the portal circulation of Chang it has been found to flow through the vessels of the upper part of the band into the portal vessels of Eng. So that the surgical anatomy of the band consists in the skin and fascia which cover it, the two separate peritoneal pouches which meet in the middle, the large peritoneal pouch, the vascular connection, to whatever extent that may exist between the two portal circulations, and the remains of the hypogastric arteries in the lower portion of the band. Thus the main difficulty in any operation for section of the band would seem to be in regard to the peritoneal processes and the portal circulation. The anastomosis which may exist between the internal mammary arteries and the intercostals in the integument in the upper portion of the band, of course would present no difficulty.

I will not venture upon any further remarks as to the surgery of the case, while there are so many distinguished gentlemen present more competent than myself to give an opinion. At the same time, operations on the peritoneum may not be considered so hazardous in this day, when ovariotomy, gastrotomy, and even Cæsarian section, are so often performed. The peritoneum-pouches themselves would not present so great a difficulty as might be anticipated, under pressure and acupuncture, by which the sensitiveness of the structure might be so altered as to permit of a section. I was informed at Mount Airy that in Paris a surgeon had made the experiment of applying pressure upon the band, and it was reported the twins had fainted in consequence. I could not ascertain, however, whether this was from fright, design, or actual pain.

As Dr. Hollingsworth is present, it may be proper for me to mention a fact which that gentleman can corroborate, that Eng was the stronger physically and Chang was the stronger mentally. The same difference was observable in their characters. Chang was more irritable than Eng, especially since an attack of paralysis with which he had been afflicted, —this being in the side next to Eng. The latter had not only to bear with the irritability of his associate, but also to support one-half his weight. Among other peculiarities, Chang would sometimes break useful articles or throw them in the fire.

In conclusion, let me say that when I turned up the skin and superficial fascia of the H incision on the posterior part of the band, I was struck with the development and the strength of the abdominal aponeuroses. The fibres arched, interlaced, and developed into a strong fibrous band about a quarter of an inch wide, running around the median line, although there was no actual joint in the cartilage.

Prof. HARRISON ALLEN:

Mr. Chairman : I will probably best discharge the duty devolving upon me by at once proceeding to a somewhat more minute anatomical description than Dr. Pancoast has given, this being in accordance with the understanding between us in reference to the evening's exercises.

Perhaps it would be best to point to that simple diagram upon the blackboard before considering the subject more fully in detail. As Dr. Pancoast has informed the Fellows, there is a union of the twins at the two ensiform cartilages, which are very firmly joined in the centre, Eng's process being the more robust of the two. You will observe that

there is a point of conjunction between the two processes which is not quite in the median line of the
band.   In the centre of the band is seen an ellip-
tical space which suggests to the observer the presence of a synovial cavity.   It is probable that the

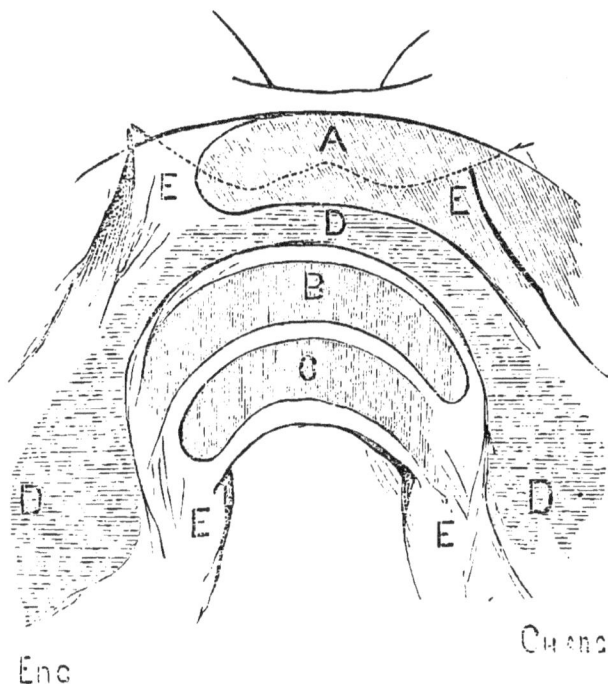

DIAGRAMMATIC REPRESENTATION OF THE BAND.

A, upper or hepatic pouch of Chang.
E, E (dotted line), union of the ensiform cartilages.
D, connecting liver band, or the " tract of portal continuity."
B, the peritoneal pouch of Eng.
C, the lower peritoneal pouch of Chang.
E, E, lower border of the band.

ensiform junction is of the character of a synchon-
drosis, with a median bursa-like sac ; neither ensiform cartilage is ossified.

Below this point, in the diagram, you see a number of differently-lined tracks. The lower one (c), immediately above the umbilicus, is only separated from the skin by a very delicate layer of tissue (so that, with the finger introduced into the pouch and moved, there is a decided indication of motion in the skin) on the under surface (e, e) of the band.

This pouch passes across the band from the abdomen of Chang, and is lost in the duplicature of the suspensory ligament of the liver of Eng. The finger passed upward to the band from the abdomen of Eng crosses the band above the pouch just mentioned, and is lost between the layers of the suspensory ligament of the liver of Chang. When the significance of the round ligament at the free border of the suspensory ligament is remembered, the relations of these pouches directly suggest that they have had essential bearings to the umbilical vein of the funis, and may be provisionally termed the *umbilical pouches.*

Above Eng's pouch (b), and between it and the under surface of the ensiform conjunction, is a second pouch (a) prolonged from Chang's abdomen, which fairly reaches the peritoneal cavity of Eng, but is not continuous with it. Extending up into this pouch from Chang's abdomen is a process which suggested to the Commission the possibility of the transit of hepatic vessels. This view was rendered more probable from the fact that a similar process passed up into the band from the liver of Eng. Accordingly, the plaster injection, colored by ultramarine, was thrown into a tributary of the portal vein of Chang, when it was observed that the fluid passed freely into the liver of Eng, as well as into some of the mesenteric veins proper. It is my own hypothesis that this bond of union (D) was the true hepatic tract ; but in its present state, in the absence of evidence of any parenchymatous admixture about the vessels thus crossing the band, we prefer to denominate the transit as the *tract of portal continuity*.

In the fœtal condition it is very likely that this large space (A), the upper pouch, now continuous with the abdomen of Chang only, was entirely occupied by true liver-tissue, which, as maturity was attained, became smaller, and left an empty space. Hence I propose to call this upper pouch the *hepatic pouch*. The contraction chanced to be greater on Chang's side, in harmony, it may be, with other evidences of a weaker and less developed type, which is so apparent in many of the tissues of Chang. Now, with reference to the demonstration. As Dr. Pancoast has already informed you, the incisions in the abdomen were made in rather an exceptional manner. By reference to the parts it will be seen that the incision in either individual was located in such a way as to avoid the median line, since it was supposed from the peculiar position of the umbilicus that the remains of the hypogastric arteries would be found extending from the fundus of the bladder upward and inward along the entire length of the anterior wall of the abdomen. Besides, this incision would enable us, by continuing from below upward, to fairly open the abdomen and examine the cord, without violating the conditions by which the Commission was bound. The flap comprises the greater part of the abdominal wall, and can be best observed, from the position of the bodies on the table, in that of Eng.   *

You notice that the tissues are well supplied with fat ; and this condition is very plainly in contrast with that of Chang.   Eng's side of the band is well nourished ; Chang's end of the band presents an entirely different aspect.   Chang was an invalid, and the weaker half of this organism, with less strength in the abdominal walls, and in every way less tissue, than was possessed by Eng.   You can mark that distinction very plainly in the two halves of the band, proving, if we had no other means of proof, that there could not be any very intimate communication of the vessels between the two.

The first point worthy of notice is that of an isolated mass of adipose tissue, evidently sub-peritoneal, which is in the position of the usual umbilicus, namely, in the median line of the abdomen, about half-way up the anterior wall.   This is strictly symmetrical, a similar point of about the same size being found in Chang.

Another fact equally well pronounced is that in Chang the bladder was found very much contracted and contained no urine ; it was deep down in the cavity of the true pelvis. That of Eng, however, was distended with urine : hence there was a

contrast in the appearance of the umbilical fold in
the two individuals, in consequence of the great
difference in the actual size of the bladders.

My finger is now in the *umbilical pouch* of Chang (C).
The motion is noticeable in the under surface of the
band. On the side of Eng no such motion will be
observed. I can very clearly see my finger passing

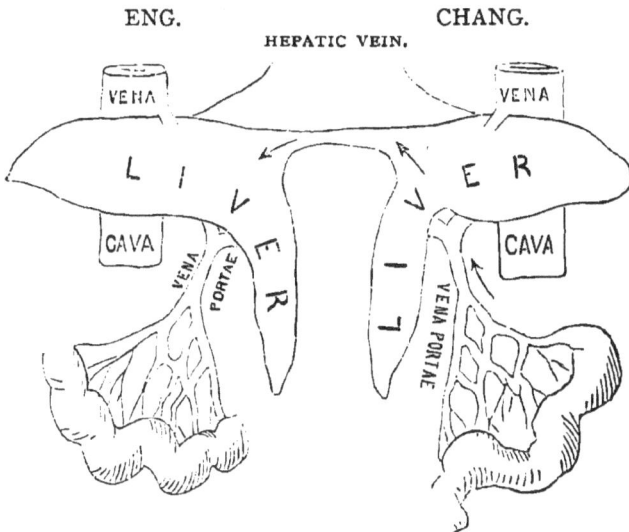

DIAGRAMMATIC REPRESENTATION OF THE LIVERS; PORTRAYING THE
RELATIONS OF THE VESSELS, ETC.

The arrows show the directions in which the injection passed from
Chang to Eng.

between the two folds of the suspensory ligament.
At this point it would perhaps be well to exhibit the
drawings which have been made of the views which
we have been able to obtain from this very limited
incision. On looking up towards the band with the
greatest possible stretch of tissue, we see the arrangement of the remains of the hypogastric arteries

2*

converging towards the bond of union. In this lower
diagram we show you the livers joined by what is
supposed to be the *tract of portal continuity.* You
will observe the limits are somewhat symmetrical.
Here is the liver of Chang, with a foreshortened
right lobe.

The remainder of the right lobe is deep within the
abdomen, and of course it has not been seen. Here
is the fundus of the gall-bladder, and there the
suspensory ligament, carrying the remains of the
umbilical vein. When the finger is passed from
Chang into Eng, it is received between the folds of
the suspensory ligament of Eng. In Eng, the parts
are essentially the same, although you see more evidence of adipose tissue. Here is a little ligament
aiding in the support of the liver, to whose con-
vexity it is attached ; it is not seen in Chang
at all. You might term it an accessory suspensory
ligament. When the finger is introduced behind
the pouch, it is observed to terminate blindly,
showing, we think, that it is adventitious, due to
the presence of that suspensory ligament.

We find some vessels of the portal system, even
as far down as the mesentery, well filled with the
blue coloring-matter. We of course desired, as
far as possible, to examine all the tissues here by
these incisions: hence it was that when the bodies
were in this position the skin was taken off from
the wall in order to get a view of the linea alba.

[The bodies were here inspected by the audience,
and afterwards turned so as to expose the posterior

part of the band. Further remarks apply to this
posterior aspect.]

Dr. PANCOAST: While the bodies are being
turned, I will take the opportunity of replying to
one or two questions which have been asked me.
First, in regard to the common sensibility of these
individuals. According to the statements we received at Mount Airy, there was a line of common
sensibility corresponding to the median line of
the band. Dr. Hollingsworth says that if a pin
were stuck into the band at the median line, both of
the twins would feel it distinctly ; but that, even at
a slight distance to either side, the point of the pin
produced an effect only on the twin of that side.

Another question has been asked me, as to whether
either of them was ever put separately under the
influence of an anæsthetic. I answer it by saying
that so far as we know it never was attempted, but
that when, upon the final occasion, Chang was
anæsthetized by death, Eng was for a time unaffected.
The story as told us at Mount Airy was that Eng
waked up and asked his son, "How is your Uncle
Chang?" The boy said, "Uncle Chang is cold—
Uncle Chang is dead." Then great excitement
took place. Eng commenced crying out immediately,—saying to his wife, whom they called in,
"My last hour is come," and finally sank away.
He was in perfect health when they went to bed.

They had been sitting up in a large double chair,
made for their accommodation. Eng was smoking

his pipe, until he became sleepy, and finally said to Chang, "We must retire." Chang said that he could not lie down comfortably. I understand that when they went from Chang's house to Eng's house [see editorial], where they died, it was against the direction of Dr. Hollingsworth; but with their usual stubbornness they persisted in riding the distance in an open buggy. To return to the narrative of the night of their death, after Chang had refused to lie down, they walked about the house for some time, and even went out to the porch, and washed their hands and drank some water. It was about one o'clock when they went to bed. Then Chang died, some time between that and morning; his death not producing any immediate impression on Eng. It was only when the latter woke up and inquired about the condition of his brother, that he was at all affected.

As to the question, "What caused Eng's death?" I am not able to tell. The post-mortem which has been made does not show the condition of his lungs. Probably the valves of his heart were in a disorganized condition, and probably also the shock upon that weakened organ caused death.

Dr. ALLEN: In my opinion, Chang died of a cerebral clot. From inquiry at his home, I was led to believe that the lung-symptoms were not due to pneumonia; indeed, were not severe enough to have been so caused. The suddenness of the death, the general atheroma of the arteries, and the fact that there had been previously an attack of cerebral paralysis, all indicated that the death was of cerebral origin. Eng probably died of fright, as the distended bladder seemed to point to a profound emotional disturbance of the nervous system, the mind remaining clear until stupor came on,—a stupor which was probably syncopal. One thing to be settled in the making of our examination was to get the bodies in the best possible position, so that we could judge of the true nature of the band.

You will observe the great contrast between the anterior appearance of the band and its posterior aspect. When we suspended them face to face we conceived we had them in the proper position for study. On the posterior side there was a fold underneath the skin extending from the ensiform cartilage of Chang, passing over, crossing the median line, and inserted into the ensiform cartilage of the opposite twin, Eng. It was one of the objects of the examination to determine what was the nature of this fold. I judge it to be the linea alba ; but I leave the Fellows to decide that for themselves. I will also add that, because we had not the privilege of cutting the anterior portion of the band, we were obliged to cut down from the point of which I have spoken to get to the structure, and demonstrate these *culs-de-sac* from behind.

Here (referring to the casts), from this point the incision is horizontal about midway, and joined laterally by two oblique lines which were directed one upward and the other downward and outward,

making a modified letter-H incision.   Thus we got
all the space we needed.   When I raise the skin, we
see the scar of the umbilicus in the superficial fascia ;
and on lifting the other flap we get a better general
demonstration.

And now we come upon the point of interest,
namely, the position of the band and its true nature.
We have a diagram here.   You notice on Chang's
side that there is an arrangement of interlacing
aponeurotic fibres, marked here ; and these fibres,
starting in Chang, pass across the median line and
are inserted into the ensiform cartilage of Eng.
Turning the lower flap downward, the upper flap
upward, and the two lateral tongues outward, the
superficial fascia is exposed.   This is abundantly
supplied with adipose tissue on either side, but is
free from fat where it covered the band.   Both the
lower flap and the fascia are lost in the scar marking
the position of the umbilicus.   The same dissection
exhibits the position of the lower pouch of Chang.
Turning down the external oblique, the two recti,
and the internal oblique muscles, the transversalis
was exposed, the latter forming a very well-defined
layer in Eng, with an interval between the ensiform
cartilage and the inferior margin of the thorax.
These were much less marked in Chang.

Turning forward this layer of fibres in Eng from
without inward, the diaphragm is brought into view.
Muscular fibres are conspicuous in this position.
The peritoneum on either side is now fairly exposed.
Incisions may now be made with a view of dem-

onstrating the pouches of the band. The upper
pouch of Chang is, you will observe, freely opened
on its posterior aspect, and the vessels in the tract
of portal continuity are seen to be well distended
with the injecting fluid. A small artery is seen
crossing beneath this tract of veins, and is probably
a branch of the hepatic; but, whatever may be its
origin, it evidently could have little effect in influ-
encing the nutrition of parts beyond the band, and
is probably retained within the band itself. The
lower pouch of Chang reveals nothing which was
not demonstrable from in front, and the same may
be said of the single pouch of Eng; thus confirm-
ing our opinions of the construction of the band
before the pouches had been opened from behind.

Dr. ABRAHAM JACOBI, of New York, being called
upon, said: I am very much obliged to the gentleman who has mentioned my name. I do not believe,
Mr. Chairman, that I have anything to add to the
stock of knowledge in regard to the subject before
us. If I were to answer the question as to how this
monstrosity originated, especially whether they became connected after having been separate organ-
isms, I should say that that idea has been given up
by those whose opinions are entitled to weight. It
is true that years ago such specimens were spoken
of by D'Alton of Halle; and a number of others
have alluded to the idea that two such individuals
might in embryonic life become united simply by
adhesion, the result of their being located together
in the embryo. In truth, it appears to me that at

that period such a thing might be possible; but of course the union would be a superficial one, not involving the deep organs.

We know that the first epidermis is formed about the end of the fifth week of embryonic life, and that after a time it is thrown off, so that the embryo of about seven or eight weeks is more loosely covered with the real epidermis than in the earlier period. The epidermis is thrown off a number of times until about the fourth month of utero-gestation, when it is finally perfected and remains intact. Now it is suggested that at those times when the epidermis is thrown off the connection takes place between the two individuals,—just as the connection takes place between the prepuce and glans, which we so often find adherent in the fœtus and in a number of new-born children.

There are evidences, which we cannot forget, that such connections have taken place before the final epidermis is formed, and about the time one of the earlier coverings is being thrown off, at a period when the internal organs, frequently implicated in such monstrosities, are already formed. There are few double monstrosities so well developed as this one. I think the records of about four hundred monsters have now been collected in the books and journals; but very few are of such a complete nature as this. Every one has heard of the Hungarian Twins, who lived to the age of twenty-one years, in the last century. Another pair of female twins, that travelled in Germany about two years ago, were described at the time, in the *Berliner Wochenschrift.* They were of a similar nature. There are two cases on record in which a division has been successfully attempted, but in those cases the connections were not so well developed as in the Siamese Twins. The connection was in the same neighborhood, but was only superficial,—of skin and subcutaneous tissue. One of the cases is recorded by Dr. Boehm (*Virchow's Archiv*). Fortunately, or unfortunately, I do not know which, they were his own children. They were of the female sex. He separated them immediately after birth. One lived three and a half days; and when the case was described in 1866, the other was five years old. In that instance the connection—three and a half inches long—extended from the ensiform process to the umbilicus. The other case is described as early as 1689, by the old German author König.

As far as the origin of twin monsters is concerned, I am certainly of those who are not of the opinion that two individuals could get into such an intimate connection by growing together. Certainly the connection is an original one. I believe that the general opinion is now that one Graafian vesicle may have two ova, or one ovum have two nuclei; and these finally may, like the two vitelli of an egg, be closed together, surrounded by the same material, forming a single complete ovum; and thus it may be that the two are included in the same ovum. I think that this will explain also why the sex is always the same,—why they are always both male or both female. They are male in twenty or twenty-five per cent. of the cases.

Dr. H. C. WOOD here asked Dr. Jacobi a question in regard to the Biddenden Sisters [an account of whom will be found in another column of this journal], as to whether they had been reported in the works on monstrosities.

Dr. JACOBI. I do not know anything about that.

Dr. PANCOAST stated that an account of those sisters was contained in a semi-popular book entitled "Lexicon Tetraglotton," published by Samuel Thomson, London, 1660.*

None of the Fellows desiring to say anything further upon the subject, on motion, the College adjourned.

ON Thursday, February 19, the Commission continued the autopsy upon the Siamese Twins, and made some important discoveries. They found that the two livers, which were supposed to be joined only by blood-vessels, were one body; the parenchymatous tissue being seemingly continuous between them.

The so-called *tract of portal continuity* is apparently liver-tissue, but the point has not yet been proven by microscopic examination. It will be

---

* The Editor of the *Times* is indebted to Dr. Pancoast for an opportunity of inspecting the work. The account of the Biddenden Maids is in a single sheet, and is evidently not a part of the original book, but has been pasted in it. In appearance it equals the body of the work in age, as shown by the color and condition of the paper; but of course it is impossible to decide with any accuracy when it was put in the book.

remembered that Chang was said to be possessed of one more pouch than Eng. When the liver was removed, however, an upper hepatic pouch was found also proceeding from Eng, so that the band contained four pouches of peritoneum, besides liver-tissue. These disclosures show that any attempt during life to separate the twins would in all probability have proved fatal.

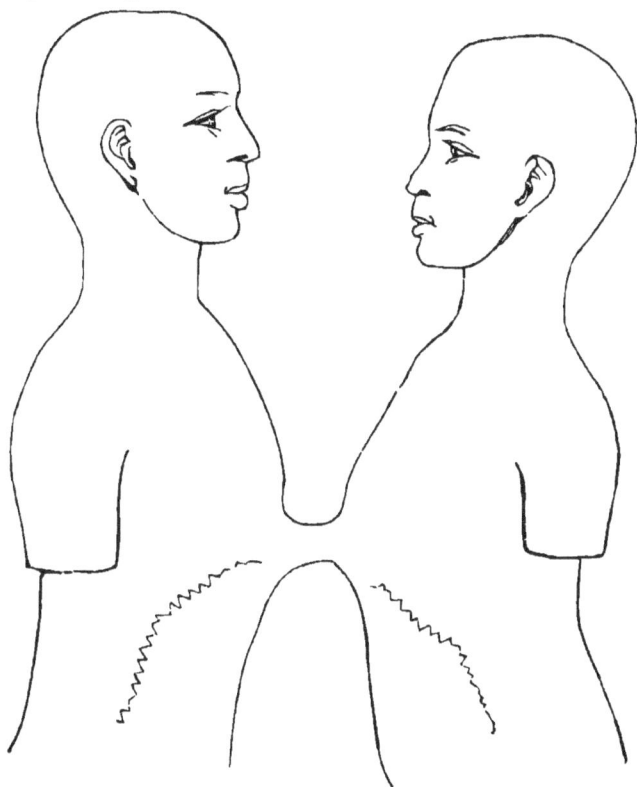

DIAGRAM FROM A CAST SHOWING THE POSITION OF THE LIGAMENT, AND OF THE PRIMARY ANTERIOR INCISIONS.

During life the twins never assumed the face-to-face position in which they are here represented, and which is without doubt that of their fœtal life.

www.ingramcontent.com/pod-product-compliance
Lightning Source LLC
Chambersburg PA
CBHW022031190326
41519CB00010B/1665